ULTIMATE SUPERCARS

MCLAREN GT

By Tamra B. Orr

Kaleidoscope
Minneapolis, MN

Bigfoot Books

The Quest for Discovery Never Ends

....

This edition first published in 2023 by Kaleidoscope Publishing, Inc.

No part of this publication may be reproduced in whole or in part without written permission of the publisher.

For information regarding permission, write to
Kaleidoscope Publishing, Inc.
6012 Blue Circle Drive
Minnetonka, MN 55343

Library of Congress Control Number
2022938004

ISBN
978-1-64519-612-9 (library bound)
978-1-64519-682-2 (ebook)

Text copyright © 2023 by Kaleidoscope Publishing, Inc. All-Star Sports, Bigfoot Books, and associated logos are trademarks and/or registered trademarks of Kaleidoscope Publishing, Inc.

Printed in the United States of America.

FIND ME IF YOU CAN!

Bigfoot lurks within one of the images in this book. It's up to you to find him!

TABLE OF CONTENTS

Chapter 1: "Ride Every Wave" 4

Chapter 2: Birth of a Race Car Driver 12

Chapter 3: A "Grand Touring Rocket" 18

Chapter 4: Waking the Car 24

Beyond the Book... 28
Research Ninja.. 29
Further Resources.. 30
Glossary.. 31
Index... 32
Photo Credits.. 32
About the Author.. 32

Chapter 1
"Ride Every Wave"

When the McLaren Company showed artist Karwai Chan their new GT, she knew just what she wanted to do. She grabbed blue and white paint. She lined up her brushes. Carefully, she began sketching crashing ocean waves. She added soaring birds. She said that she was inspired by how "fearless and unstoppable" a McLaren engine is—just like waves. Chan spent 200 hours on the painting. She named it "Ride Every Wave."

For years, McLaren has produced fast, sleek race cars. In 2019, it added the GT model to its lineup. This McLaren is not for racetracks. It's for the open road.

One look at a McLaren GT and it is clear it goes fast. It can reach 124 miles per hour (199 km/h) in only nine seconds. This model is different for McLaren. It is not designed to win races! It is designed for anything from running errands to going on road trips. Storage space is big enough to hold a suitcase, a pair of snow skis, or a set of golf clubs. The seats are soft leather.

After opening the GT's scissor doors, owners flick the start button. This engine does not roar like a race car. It purrs.

PARTS OF A
MCLAREN GT

front trunk

rounded nose

streamlined headlights

One of the GT's best features for long trips is its glass wraparound roof. It provides a **panoramic** view. The driver and passengers can see in every direction, including overhead. It is like driving down the road in a clear, fast-moving bubble! A button on the ceiling allows drivers to dim the lights in case the sunshine is too bright.

A seven-inch (18 cm) touchscreen on the McLaren GT's dashboard allows drivers to control the music system. They can also turn the car on and off. They can even use the **GPS** system, all with just the touch of a finger.

WIDE AND COMFORTABLE

McLaren's race cars have narrow, thin seats. They are often uncomfortable because drivers are not sitting on them for very long. The GT seats are wider and thicker for the driver who plans to be on the road for hours.

FUN FACT
McLaren's fastest car is the Speedtail. It can reach speeds of 250 miles per hour (402 km/h).

Chapter 2
Birth of a Race Car Driver

Almost from the day he was born in New Zealand in 1937, Bruce McLaren loved cars. He helped at his family's service station. He knew all the mechanics. Before long, he began riding and racing motorcycles. His father, Les, helped him build his first race car. By the age of 15, Bruce was competing in races. He studied engineering in college. Soon, he was off to Europe to race with other professional drivers.

In 1959, McLaren won the U.S. Grand Prix at only 22! Six years later, he created his own race team. McLaren was a driver and designer.

In 1970, McLaren was testing one of his cars on a racetrack in England. The rear bodywork of his car suddenly came off at 170 miles per hour (273 km/h). McLaren crashed. He was killed instantly. He was only 32 years old.

Chris Amon, one of his friends, stated that McLaren was a wonderful person. "You never knew what he was going to come up with the next day," said Amon. "He had flashes of inspiration, and having decided on something, he never allowed anything to get in the way of getting it done."

WHERE THE MCLAREN GT IS MADE

Woking, England: McLaren Technology Centre

KEEPING IT COOL

The McLaren headquarters are in Surrey, England at the McLaren Technology Centre. The property has four ecology lakes that are used as heat exchangers to cool the building.

McLaren was gone. But his company went on. It is still making fast cars today. Not only does it make race cars, but it makes high-performance cars for daily use. The first car for the street was launched in 1992.

FUN FACT

Many sports cars hug the ground, but the McLaren GT has 4.3 inches (10.9 cm) clearance to make it more comfortable.

Almost 20 years later, McLaren produced the MP4-12C with a 3.8-liter V8 engine behind the driver, instead of in front. This car could go from zero to 60 miles per hour (96 km/h) in only 2.8 seconds. Since then, McLaren has produced the P1, 650S, 570S, and now, the GT. The company has plans to release at least 18 other designs in the coming years.

PILING ON THE EXTRAS

The base price for a McLaren GT is $200,000. With extras like special floor mats, seat covers, cameras, shift knobs, chrome trim, or custom wheels, it can cost over $240,000.

Chapter 3
A "Grand Touring Rocket"

When Tom Voelk, a well-known car reviewer for *Driver*, took a McLaren GT out for a test drive, he was amazed. "It's like an alien craft visiting from a world that worships art and sculpture," he said. "It looks stunning in motion and it looks stunning standing still." After his drive, he stated this car was a "grand touring rocket."

There are a lot of ways to make the car look unique too. Anyone buying a McLaren GT can choose from dozens of paint colors. There are nine colors available for the brake **calipers**. Even the seat stitching comes in five different colors.

THE MCLAREN GT IN DETAIL

COST: $200,000 basic model

Height: doors closed 4 feet (1.2 m) doors open 6.5 feet (1.9 m)

Width: 6.9 feet (2.1 m)

LENGTH: 15.4 feet (4.7 m)

WEIGHT: 3,384 pounds (1,534 kg)

TOP SPEED: 203 miles per hour (326 km/h)

TIME FROM 0 to 60 miles per hour (96 km/h): 3.1 seconds

Although the McLaren GT is for road trips, it is still one of the world's most powerful cars. With its 4-liter, **turbocharged** V8 motor, it packs 612 **horsepower**. The GT's carbon-fiber **chassis** makes the car strong but lightweight. Add in a **hydraulic** steering system and steel brakes, and the GT needs only a finger to steer and a toe to stop.

The GT model is a mid-engine car. It is behind the front seat but in front of the rear axle.

There are two areas for storage. The first is in the front. It is sometimes called a "frunk."

FUN FACT
Frunk is made from the words *front* and *trunk*.

Unlike most fast cars, the McLaren GT has a second storage spot right over the engine. It is big enough to pack a few bags. However, since it can get hot next to an engine, it is not a good place to put a cooler or a box of chocolate.

A MINI-GT

For their youngest fans, McLaren created four junior ride-ons. The GT model features scissor doors. When young drivers turn the key, they can hear engine noises. Pushing on the gas pedal creates a revving noise. When they push the brakes, LED taillights come on.

Chapter 4
Waking the Car

Imagine sliding in across the McLaren's leather seats. You fasten the seatbelt. You adjust the mirrors. Next, you wake up the GT. Hit the start button. Tap the brakes. Push the button one more time. Listen to that engine rumble. Finally, it is time to roll—or maybe rocket.

GT drivers can choose between driving in automatic or in manual **transmission** as the car has a 7-speed dual-clutch **gearbox**. Shifting through the gears only takes the touch of a thumb thanks to paddle shifters, small levers on the steering column. One light touch is all it takes. No clutch pedal is needed!

SUPER FABRIC

Even the material used inside a GT is extra high-tech. It resists stains and splashes. It is super tough. They make the fabric from layers of tiny, armored plates that make it almost impossible to absorb a liquid or to be cut or scraped.

It is time to gobble up the road in the McLaren GT. Listen to the engine's powerful growl as it eats up mile after mile. Enjoy the smooth, easy turns. Feel the quick stops. Watch the scenery passing in every direction. Then, at the end of the day, come back home, stop, turn off the car, and let it rest.

FUN FACT
The McLaren GT has three driving styles: comfort, sport, and track. Each one changes how the car moves and runs.

For many years, McLaren Company has designed cars that thrill its drivers. Whether it is a race car driver hoping to come in first or a personal owner who wants an exotic ride along country back roads, McLaren has something to offer.

BEYOND
THE BOOK

After reading the book, it's time to think about what you learned. Try the following exercises to jump-start your ideas.

RESEARCH

THAT'S NEWS TO ME. In the second chapter, Bruce McLaren won the U.S. Grand Prix at only 22. How might news sources be able to fill in more detail about this? What new information could you find in news articles? Where could you go to find those sources?

CREATE

SHARPEN YOUR RESEARCH SKILLS. The McLaren GT has a dual-clutch gearbox. Where could you go in the library to find more information about dual-clutch gearboxes? Who could you talk to who might know more? Create a research plan. Write a paragraph about your next steps.

SHARE

SUM IT UP. Write one paragraph summarizing the important points from this book. Make sure it's in your own words. Don't just copy what is in the text. Share the paragraph with a classmate. Does your classmate have any comments about the summary? Do they have additional questions about the McLaren GT?

GROW

REAL-LIFE RESEARCH. What places could you visit to learn more about cars like the McLaren GT? What other things could you learn while you were there?

RESEARCH NINJA

Visit www.ninjaresearcher.com/6129 to learn how to take your research skills and book report writing to the next level!

RESEARCH

DIGITAL LITERACY TOOLS

SEARCH LIKE A PRO
Learn about how to use search engines to find useful websites.

FACT OR FAKE?
Discover how you can tell a trusted website from an untrustworthy resource.

TEXT DETECTIVE
Explore how to zero in on the information you need most.

SHOW YOUR WORK
Research responsibly—learn how to cite sources.

WRITE

GET TO THE POINT
Learn how to express your main ideas.

PLAN OF ATTACK
Learn prewriting exercises and create an outline.

DOWNLOADABLE REPORT FORMS

Further Resources

BOOKS

Baby Professor. *How are Sportscars Made?* Nova Scotia: Baby Professor, 2015.

Mason, Paul. *British Supercars: McLaren, Aston Martin, Jaguar.* New York: PowerKids Press, 2018.

Myers, Carrie. *McLaren 12C (Ultimate Supercars).* Minnesota: North Star Editions, 2019.

WEBSITES

Factsurfer.com gives you a safe, fun way to find more information.

1. Go to www.factsurfer.com.
2. Enter "McLaren GT" into the search box and click 🔍
3. Select your book cover to see a list of related websites.

Glossary

axle: a shaft on which a wheel(s) turn.

chassis: frame of a motor vehicle.

gearbox: a set of gears with its casing.

GPS: Global Positioning System used as a worldwide navigational tool based on satellites.

GT: this stands for "Grand Tourer." It means a blend of a sports car and a luxury model car.

horsepower: the measure of the power of an engine.

hydraulic: operated by means of water.

panoramic: a wide view surrounding a person on all sides.

transmission: the mechanism by which power is transmitted from an engine to the wheels of a motor vehicle.

turbocharged: an engine that is driven by the engine's exhaust gases.

Index

Bruce McLaren, 12
company, 4, 16, 17, 27
driver, 10, 12, 13, 17, 25, 27
engine, 4, 6, 17, 22, 23, 24, 26
gear, 25
GT, 4, 6, 10, 17, 18, 21, 22, 23, 24, 25, 26

model, 4, 6, 22
powerful, 21, 26
race car, 4, 6, 12, 16, 27
racetrack, 14
road trip, 6, 21
scissor door, 6
seat, 6, 18, 22, 24
wave, 4

PHOTO CREDITS

The images in this book are reproduced through: Jack Skeens/Shutterstock 8; All other images courtesy of McLaren Press Room (Chris Brown 14).
Cover: Courtesy of McLaren Press Room, YIUCHEUNG/Shutterstock (background).

About the Author

Tamra B. Orr is a full-time author living in the Pacific Northwest with her family. She attended Ball State University before moving cross-country. Orr has written more than 750 books for readers of all ages and says it is the best job in the world because she is always learning something new.